EXPLORING
COMETS

The Rosen Publishing Group's
PowerKids Press™
New York

JENNIFER WAY

Published in 2007 by The Rosen Publishing Group, Inc.
29 East 21st Street, New York, NY 10010

First Edition

Editor: Amelie von Zumbusch
Book Design: Ginny Chu

Photo Credits: Cover, p. 6 © Shigemi Numazawa / Atlas Photo Bank / Photo Researchers, Inc. p. 4 © John Chumack / Photo Researchers, Inc.; p. 8 © European space agency / Photo Researchers, Inc.; p. 10 © Mark Garlick / Photo Researchers, Inc.; p. 12 © Eckhard Slawik / Photo Researchers, Inc.; p.14 © Jerry Schad / Photo Researchers, Inc.; p. 16 © Royal Greenwich Observatory / Photo Researchers, Inc.; p. 18 © Claus Lunau / FOCI / Bonnier Pub. / Photo Researchers, Inc.; p. 20 NASA/JPL/UMD

Library of Congress Cataloging-in-Publication Data

Way, Jennifer.
  Exploring comets / Jennifer Way.— 1st ed.
     p. cm. — (Objects in the sky)
  Includes bibliographical references and index.
  ISBN 1-4042-3469-1 (lib. bdg.) — ISBN 1-4042-2177-8 (pbk.)
  1. Comets—Juvenile literature. I. Title. II. Series.
  QB721.5.W39 2007
  523.6—dc22
                                                2006000736

Manufactured in the United States of America

Contents

The comet Hale-Bopp, shown here, was discovered by Alan Hale and Thomas Bopp in 1995.

What Is a Comet?

Have you heard of Halley's comet or the comet Hale-Bopp? They are two comets that have passed Earth in the last few years. Comets are pieces of rock, ice, and **frozen** gases that orbit, or circle, the Sun.

Throughout history people have recorded seeing comets. Around 300 years ago, **astronomers** discovered that comets return after a set number of years. Since that time astronomers have learned much about comets. They believe that comets were formed around the same time as our **solar system**.

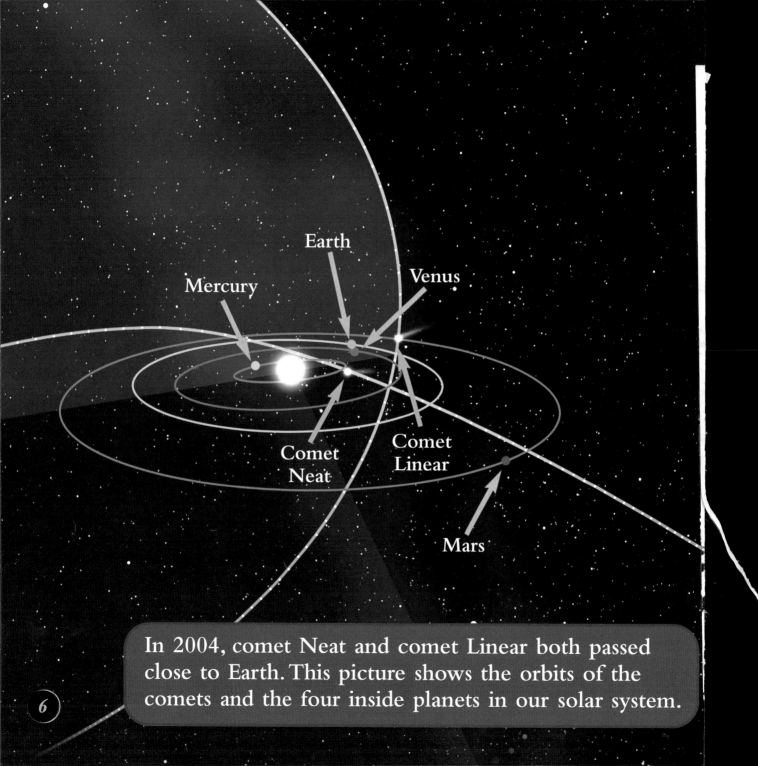

Mercury

Earth

Venus

Comet
Neat

Comet
Linear

Mars

In 2004, comet Neat and comet Linear both passed close to Earth. This picture shows the orbits of the comets and the four inside planets in our solar system.

Comets travel in paths called orbits. These orbits are caused by the Sun's **gravity**. The gravity of the **planets** a comet passes also helps shape its orbit. Comet orbits are highly elliptical. This means that the shapes of their orbits are long ovals.

The size of a comet's orbit sets how long that comet takes to move around the Sun. A comet's orbit can take as little as three years or as long as millions of years! We see comets only when their orbits cross near Earth's orbit.

7

This picture of the nucleus of Halley's comet was taken in 1986. The comet's nucleus is about 9.5 miles (15.3 km) long and 4.3 miles (7 km) wide.

The Nucleus

The **nucleus** is the only part of a comet that is always present. It can be thought of as a sort of dirty snowball. The nucleus is made up of small pieces of rock, ice, and frozen gases. Some of the frozen gases found in the nucleus are **water vapor**, **carbon monoxide**, and **carbon dioxide**. The core, or center, of the nucleus sometimes has a small rock around which the rest of the nucleus has frozen. The nucleus is the smallest part of a comet. The nucleus of a middle-sized comet is only about 6 miles (10 km) across.

A comet's coma can only be seen when the comet's orbit brings it close to the Sun.

THE COMA

The **coma** is a cloud of gas that circles the comet's nucleus. The coma is larger than the nucleus. It is most often about 62,137 miles (100,000 km) around. The coma and the nucleus together make up the comet's head.

The coma appears only when the comet nears the Sun. As the Sun warms the comet, the ice on the outside of the nucleus turns into gas. This is called sublimation. The gases glow and can be seen from Earth. As the gases continue to warm, the coma grows larger.

You can see the dust tail in this picture of comet Hyakutake. A dust tail can be millions of miles (km) long.

Most comets have a yellowish dust tail. The dust tail forms as the **solar wind** pushes dust in the coma away from the rest of the comet. The dust tail bends because of the comet's elliptical orbit.

When the tail first forms, it points toward the Sun. As the comet gets closer to the Sun, the solar wind gets stronger. It pushes the dust tail away from the Sun. Sunlight shines back the dust, which can make it possible to see the tail from Earth. As the comet moves away from the Sun, the dust tail begins to fade.

You can see the blue ion tail of comet Hale-Bopp in this picture. The comet's dust tail looks white.

The Ion Tail

A comet's **ion** tail is longer than its dust tail. An ion tail can be up to 100 million miles (161 million km) long. The ion tail forms as the comet nears the Sun. During sublimation some of the ice becomes ionized. This means that when the ice turns to gas, some of that gas becomes charged. These charged gases are what cause the ion tail to glow blue.

The solar wind pushes the ion tail away from the Sun in a line. Like the dust tail, the ion tail begins to fade as it moves away from the Sun.

Halley's comet, shown here, is a short-period comet. Its orbit takes between 74 and 79 years.

Short-period Comets

Astronomers group comets based on how long their orbits take. Short-period comets have orbits of fewer than 200 years.

Many astronomers believe that short-period comets come from the Kuiper Belt. The Kuiper Belt is a disk-shaped belt of matter beyond the orbit of the planet Neptune. Astronomers think it is filled with small, icy objects. They believe that the pull of gravity from nearby planets and crashes with other space matter cause comets to leave the Kuiper Belt. They then begin their elliptical orbit around the Sun.

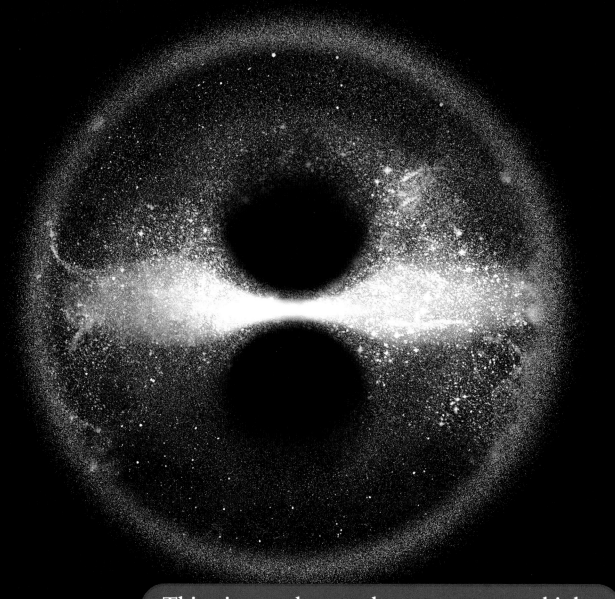

This picture shows what astronomers think the Oort Cloud might look like. The Kuiper Belt is the bright spot across the middle.

Long-period Comets

Astronomers call comets whose orbits take longer than 200 years long-period comets. The comet Hale-Bopp is a long-period comet with an orbit of about 2,380 years. Some long-period comets have orbits that take 30 million years!

Long-period comets are believed to come from the Oort Cloud. This round cloud of matter lies far past the Kuiper Belt. Astronomers believe the Oort Cloud is made of dust and ice. They think that objects in the Oort Cloud become comets when passing stars push them into our solar system.

In 2005, astronomers sent a space probe to crash into the comet Tempel 1. This drawing shows a spacecraft about to send out the probe on July 4, 2005.

Studying Comets

Astronomers study comets using telescopes. Telescopes are tools that let us see faraway things by **magnifying** them. Some telescopes are small enough to hold in your hand. Astronomers use large telescopes in buildings called observatories. These telescopes let them see into outer space.

Space probes are another tool astronomers use to study comets. A space probe is a machine that is sent into space to take pictures and record facts. It allows astronomers to look more closely at comets than they can from Earth.

Comets Are Cool!

Comets do not last for all time. Most comets break up over time. Some comets crash into planets or the Sun and are destroyed. Comets that crash into the Sun are called sungrazers.

Comets are among the oldest parts of our solar system. Astronomers believe comets are icy leftovers from the beginning of our solar system, more than four **billion** years ago. By studying comets astronomers hope to learn more about how our planet and solar system began.

Glossary

astronomers (uh-STRAH-nuh-merz) People who study the Sun, the Moon, the planets, and the stars.

billion (BIL-yun) One thousand millions.

carbon dioxide (KAR-bin dy-OK-syd) A colorless gas. People breathe out carbon dioxide.

carbon monoxide (KAR-bin muh-NOK-syd) A colorless gas that is unsafe to breathe.

coma (KOH-muh) The cloud of gas around a comet's center.

frozen (FROH-zn) Hard and very cold.

gravity (GRA-vih-tee) The force that pulls two objects together.

ion (EYE-un) A tiny piece of charged matter.

magnifying (MAG-nuh-fy-ing) Making an object appear larger than it is.

nucleus (NOO-klee-us) The center of something.

planets (PLA-nets) Large objects, such as Earth, that move around the Sun.

solar system (SOH-ler SIS-tem) A group of planets and other objects that circle a star.

solar wind (SOH-ler WIND) Tiny pieces of charged matter that flow from the Sun.

water vapor (WAH-ter VAY-pur) The gaseous state of water.

Index

A
astronomers, 5, 17, 19, 21–22

C
carbon dioxide, 9
carbon monoxide, 9
comet Hale-Bopp, 5, 19

D
dust tail, 13, 15

G
gravity, 7, 17

H
Halley's comet, 5

K
Kuiper Belt, 17, 19

N
Neptune, 17
nucleus, 9, 11

O
observatories, 21
Oort Cloud, 19
orbit(s), 7, 13, 17, 19

S
solar system, 5, 19, 22
solar wind, 13, 15
space probes, 21
sublimation, 11, 15
sungrazers, 22

T
telescopes, 21

W
water vapor, 9

Web Sites

Due to the changing nature of Internet links, PowerKids Press has developed an online list of Web sites related to the subject of this book. This site is updated regularly. Please use this link to access the list:
www.powerkidslinks.com/oits/comets/